Selling Circuit Boards for Profit Guide

Roggie Clark

Continue Learning At Our Website: Roggieclarkbooks.com

Continue exploring educational resources, tools, books, workbooks, and learning materials designed to help readers continue building knowledge and practical skills. Visit: www.roggieclarkbooks.com

Inside you'll find:

- Educational worksheets
- Beginner-friendly guides
- Printable activities
- Companion materials for books and courses

A growing library of educational content for readers, collectors, learners, parents, teachers, and lifelong students.

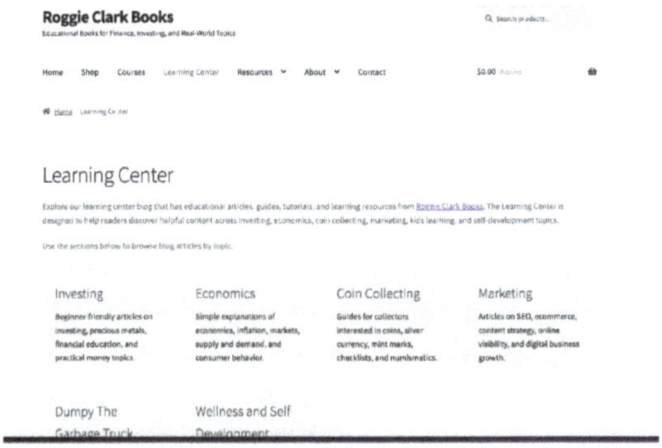

Learn Macro and Micro Economics, With This One Book

Whether you're a student, a professional, or simply someone eager to understand the world around you, this book is your perfect introduction to the fundamentals of micro and macroeconomics.

https://www.amazon.com/Macro-Micro-Economics-Workbook-Beginners-ebook/dp/B0FVPMFX6B

The Strategic Guide to U.S. Mint Gold and Silver Coin Investing

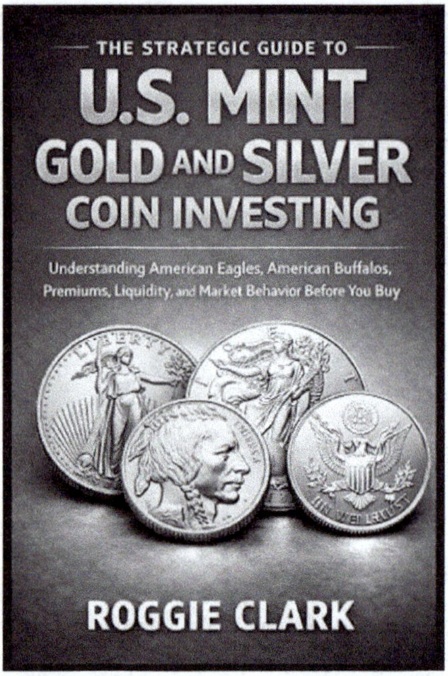

Rather than offering a general introduction to precious metals, this guide examines how American Gold Eagles, American Gold Buffalos, and Silver American Eagles behave in real investor markets.

https://www.amazon.com/Strategic-Guide-U-S-Silver-Investing/dp/B0GQKJF9K6

Copyright

Title: Selling Circuit Boards for Profit Guide
Author: Roggie Clark

Cover Design: Roggie Clark
Editing: Renee' Clark

Printed in the United States of America

Legal Disclaimer

The information presented in this book is provided solely for general educational and informational purposes. It is intended to help readers better understand circuit board recycling, electronic waste (e-waste), and responsible sourcing practices. It should not be interpreted as legal, financial, tax, environmental, engineering, electrical, recycling, or professional business advice.

The author and publisher make every effort to provide accurate and current information. However, laws, regulations, environmental requirements, recycling policies, data privacy obligations, auction procedures, electronic waste regulations, and buyer acceptance standards vary by country, state, province, municipality, and organization, and they may change over time. Readers are responsible for researching and complying with all applicable laws, regulations, and requirements within their own jurisdiction.

This book promotes only **legal and ethical methods** of obtaining and selling circuit boards and electronic waste. Nothing contained in this publication should be interpreted as encouraging theft, trespassing, unauthorized access to property, improper disposal of electronic waste, unlawful collection of electronic equipment, violation of data privacy laws, or any other illegal or unethical activity. Always obtain permission before collecting, purchasing, transporting, dismantling, or selling electronic equipment that belongs to another individual, business, or organization.

Many electronic devices contain potentially hazardous components, including batteries, capacitors, sharp metal edges, glass, chemicals, and electrical components that may present safety risks if handled improperly. Readers should always use appropriate personal protective equipment, follow recommended safety procedures, and comply with all applicable environmental and workplace safety regulations. Never dismantle energized electrical equipment or devices that present a risk of electrical shock or other hazards.

Many computers, servers, networking devices, storage systems, and other electronic products may contain sensitive or confidential information. Readers are responsible for complying with all applicable privacy laws, data protection requirements, and organizational policies when handling electronic equipment. Whenever appropriate, electronic storage devices should be securely managed through approved data destruction or certified data sanitization procedures before reuse or recycling.

References to auctions, recycling companies, businesses, organizations, manufacturers, government agencies, products, services, or electronic equipment are provided for educational purposes only and do not constitute endorsements or recommendations. Availability, pricing, grading standards, purchasing policies, and recycling procedures vary among organizations and may change without notice.

Any discussion of potential profits, income, resale value, or business opportunities is intended for illustrative purposes only and should not be interpreted as a guarantee of financial success. Individual results will vary depending on factors including experience, sourcing methods, market conditions, operating costs, buyer demand, equipment availability, and individual effort.

By using the information contained in this book, you acknowledge that you assume full responsibility for your own decisions and actions. The author and publisher disclaim all liability for any loss, injury, damage, financial loss, legal consequences, environmental liability, or other claims arising directly or indirectly from the use or misuse of the information presented in this publication.

Image Credits and Public Domain Notice

All images, illustrations, graphics, photographs, visual references, and related materials included in this publication are believed to be in the public domain, used under applicable licenses or permissions, created by the author or publisher, or obtained from sources that clearly identify them as available for lawful use.

The author and publisher have made a good faith effort to use only visual materials that are lawfully available for use and distribution. Where applicable, images and visual materials may have been resized, cropped, edited, enhanced, stylized, modified, reformatted, or arranged for educational, informational, design, formatting, or presentation purposes.

This publication is intended for educational, informational, and illustrative purposes only. The inclusion of any image, artwork, photograph, visual reference, symbol, historical material, design, or graphic element is not intended to claim ownership over public domain materials or any third-party intellectual property rights that remain with their respective owners.

If you believe any image or visual material included in this publication has been used in error, please contact the author or publisher so the matter may be reviewed and corrected in future editions or releases if necessary.

Any factual information, specifications, historical references, values, descriptions, statistics, or related content accompanying visual materials are provided for general informational purposes only and may change over time. Readers should conduct their own independent research and consult appropriate sources or qualified professionals when making decisions based on information presented in this publication.

This publication is independently produced and is not affiliated with, endorsed by, sponsored by, or associated with any company, organization, government agency, institution, trademark holder, or third party unless specifically stated otherwise.

Table of Contents

Introduction

Electronic devices have become part of nearly every aspect of modern life. Computers, smartphones, servers, networking equipment, televisions, industrial controls, gaming systems, medical equipment, and countless other technologies all rely on **circuit boards** to function. As these devices become outdated or reach the end of their useful lives, many are replaced and discarded. While most people see electronic waste, experienced recyclers recognize that many of these circuit boards still have significant value. Understanding how to identify, source, grade, and sell them legally has become an increasingly popular opportunity for hobbyists, entrepreneurs, and anyone interested in the growing world of electronics recycling.

Why Circuit Boards Have Become So Valuable

The rapid pace of technology means that millions of electronic devices are retired every year. Businesses upgrade their computer systems, schools replace aging equipment, manufacturers modernize production lines, and consumers purchase newer electronics. Although these products may no longer serve their original purpose, many still contain valuable materials that can be recovered through responsible recycling.

Circuit boards are valuable because they contain a combination of useful metals and electronic components. Depending on the type of board, they may contain copper, gold, silver, palladium, aluminum, and other materials that specialized recycling companies recover and reuse. Not every circuit board contains the same amount of valuable material, however. Some boards command significantly higher prices because they contain larger quantities of recoverable metals or come from specialized industrial equipment.

Social media and online marketplaces have also contributed to the growing interest in circuit board recycling. Videos showing pallets of discarded electronics, bins of computer motherboards, and successful e-waste businesses have inspired many people to ask the same question: "Where do these recyclers find all of these circuit boards?" The answer is that most successful recyclers build relationships with legitimate sources and understand how electronic equipment naturally enters the recycling stream.

As you'll discover throughout this book, the real opportunity is not simply collecting circuit boards. It is learning how to recognize valuable boards, understand their grading, and consistently source them through ethical and legal methods.

How This Book Will Help You

Beginning in the electronic recycling industry can feel overwhelming. There are many different types of circuit boards, numerous grading systems, changing market prices, and countless opinions shared online. Some videos focus only on dramatic profits, while others encourage processing methods that may be unsafe, environmentally harmful, or even illegal.

This guide takes a different approach.

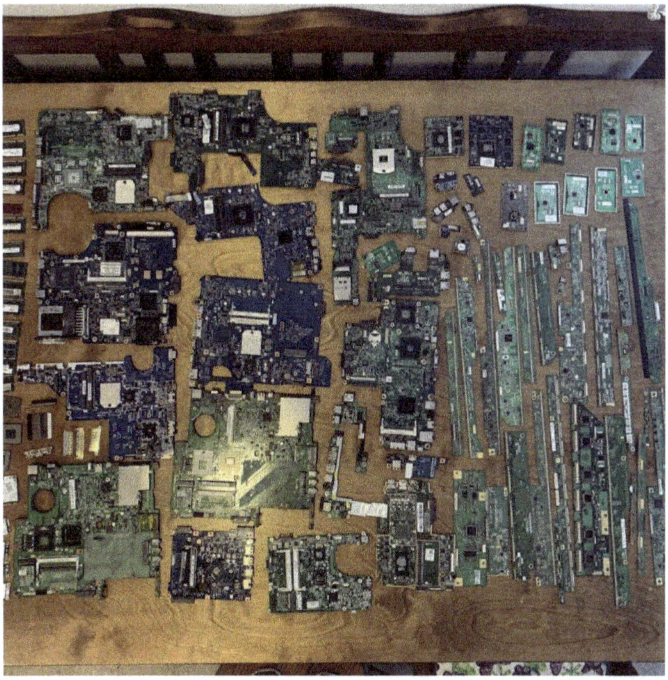

Rather than focusing on recovering precious metals yourself, this book explains how to participate in the electronic recycling industry by sourcing and selling circuit boards to professional buyers. You'll learn how different boards are classified, where they can be legally obtained, how electronic scrap buyers evaluate them, and how to build a practical side business based on responsible sourcing and informed decision-making.

Each chapter builds upon the previous one, helping you develop the knowledge needed to identify valuable opportunities while avoiding many of the mistakes that discourage beginners. Whether your goal is earning extra income, reducing electronic waste, expanding your recycling knowledge, or starting a small business, this guide is designed to provide practical information you can apply immediately.

Ethical, Legal, and Safety First

One of the most important principles discussed throughout this book is that every circuit board should be obtained legally and ethically. Businesses, schools, government agencies, manufacturers, homeowners, and recycling facilities all replace electronic equipment on a regular basis, creating legitimate opportunities to acquire electronic scrap through auctions, surplus sales, IT asset disposals, business liquidations, electronics recycling programs, and other authorized channels.

METAL
PLASTIC
BATTERY

Electronic recycling also requires attention to safety. Some devices may contain sharp edges, batteries, capacitors, glass components, or other materials that require careful handling. Certain equipment should never be dismantled without proper knowledge or safety precautions. Throughout this guide, the focus remains on responsible sourcing, safe handling practices, and respecting all applicable laws and property rights.

Operating with integrity benefits everyone involved. Businesses are more likely to work with trustworthy recyclers, buyers prefer well-organized and honestly represented materials, and responsible recycling helps reduce unnecessary electronic waste while supporting the recovery of valuable resources.

Last Thoughts

The electronic recycling industry continues to grow as technology evolves and more devices reach the end of their useful lives. By understanding why circuit boards have value, how this book is designed to help you, and why ethical practices are essential, you'll be well prepared to begin exploring this exciting field. In the next chapter, you'll learn the fundamentals of circuit boards and e-waste, including the different types of boards you'll encounter, why some are more valuable than others, and how grading systems help determine their market value.

Chapter 1: Understanding Circuit Boards and E-Waste

Before you begin sourcing or selling circuit boards, it's important to understand exactly what they are and why they have become valuable within the recycling industry. Not every circuit board is the same, and not every board commands the same price. Learning to recognize different types of boards, understand how buyers classify them, and identify the characteristics that influence value will help you make better purchasing and sourcing decisions from the very beginning. Like any successful recycling business, knowledge is one of your most valuable assets.

What Circuit Boards Are and Why They Have Value

A circuit board serves as the foundation of nearly every electronic device. It provides the pathways that allow electricity to travel between processors, memory chips, connectors, sensors, and countless other electronic components. Without circuit boards, modern computers, televisions, smartphones, industrial equipment, communication systems, medical devices, and household appliances would not function.

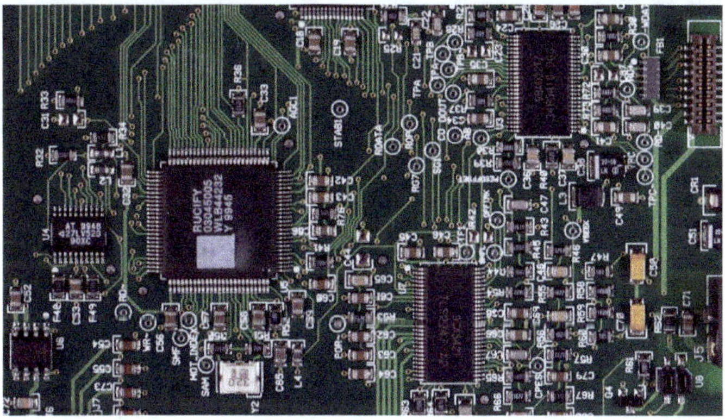

Although circuit boards appear to be simple green, blue, or brown panels covered with electronic components, they are actually manufactured using a combination of metals, fiberglass, plastics, and specialized materials. Copper forms many of the electrical pathways found within the board itself, while connectors, processors, and other components may contain small amounts of precious metals such as gold, silver, and palladium. These materials help improve electrical conductivity, resist corrosion, and increase long-term reliability.

The value of a circuit board is rarely determined by appearance alone. Instead, buyers consider factors such as the type of equipment the board came from, the amount of recoverable material it contains, the density of electronic components, and the demand for that particular category of electronic scrap. A large circuit board is not automatically worth more than a smaller one, and older equipment is not always more valuable than newer technology.

Understanding this principle helps explain why experienced recyclers spend time identifying boards before selling them. The more accurately you recognize different categories of electronic scrap, the more confidently you can evaluate opportunities when sourcing materials.

Common Types of Circuit Boards

One of the first skills you'll develop is recognizing the many different types of circuit boards that enter the recycling market. While there are thousands of designs, most beginner recyclers quickly become familiar with several common categories that appear regularly in homes, businesses, schools, and industrial facilities.

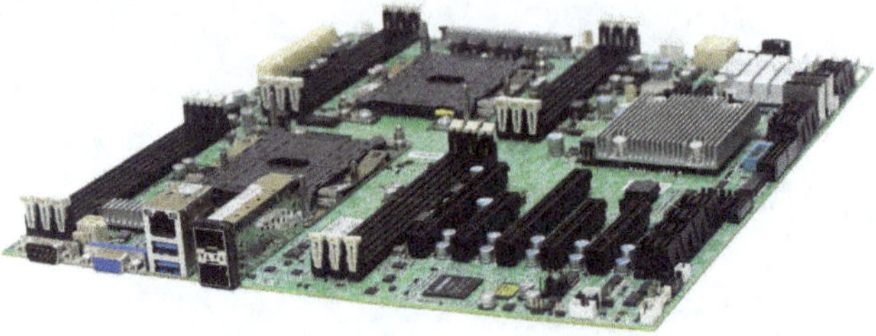

Computer motherboards are among the most recognizable circuit boards because they serve as the primary control board inside desktop computers and servers. Expansion cards, graphics cards, memory modules, and processor boards are also frequently recovered from computer equipment during upgrades or electronic recycling projects.

Networking equipment represents another important category. Routers, switches, modems, wireless access points, and telecommunications equipment often contain boards with valuable electronic components. Businesses replacing office networks or upgrading communication systems may generate significant quantities of these materials during equipment refresh cycles.

Industrial equipment, medical devices, manufacturing controls, security systems, point-of-sale terminals, laboratory equipment, and commercial electronics also contain specialized circuit boards. Although these boards may look unfamiliar to beginners, they often become valuable opportunities because they are produced for professional applications rather than consumer use.

Consumer electronics should not be overlooked either. Televisions, DVD players, gaming consoles, printers, stereos, power supplies, microwaves, washing machines, and many other household devices contain circuit boards that eventually enter the electronic recycling stream. While some may have modest value individually, larger quantities can become worthwhile when sourced consistently.

As your experience grows, you'll begin recognizing circuit boards almost immediately, regardless of the equipment they came from. This ability develops naturally through observation and repeated exposure to different types of electronic devices.

Understanding Circuit Board Grades

Just as scrap copper is sorted into different grades, circuit boards are also classified according to their overall value and the materials they contain. These grading systems help electronic scrap buyers determine pricing while making it easier to sort large quantities of incoming material.

Although grading systems vary somewhat between buyers, most electronic recyclers separate boards into broad categories based on their expected recovery value. Boards containing larger numbers of processors, integrated circuits, gold-plated connectors, and densely populated electronic components are generally considered more valuable than boards containing relatively few components.

Server boards, telecommunications boards, certain industrial control boards, and high-performance computer boards often receive higher grades because they typically contain greater concentrations of recoverable materials. Simpler boards removed from household appliances or consumer electronics may be assigned lower grades, even though they still retain recycling value.

One of the biggest mistakes beginners make is assuming every circuit board should be sold together. Mixing higher-value boards with lower-grade material may reduce the overall value of an entire shipment. Experienced recyclers usually separate boards according to type before contacting buyers, allowing each category to be evaluated independently.

It is also important to remember that grading standards are not universal. One buyer may classify a particular board differently than another depending on processing methods, refining capabilities, and current market demand. Developing relationships with multiple electronic scrap buyers and understanding their grading systems will help you make more informed selling decisions over time.

The more familiar you become with circuit board grading, the easier it becomes to estimate potential value before purchasing or collecting electronic scrap. That knowledge helps you spend your time and resources more efficiently while reducing costly mistakes.

Last Thoughts

Understanding circuit boards is the foundation of every successful electronic recycling operation. By learning what circuit boards are, recognizing the most common categories, and understanding how buyers classify them, you'll be better prepared to evaluate opportunities with confidence. In the next chapter, you'll shift your attention from identifying circuit boards to discovering the many legal and ethical places where they can be sourced, from household electronics and business upgrades to public auctions, IT asset disposals, and other legitimate opportunities.

Chapter 2: Finding Circuit Boards Legally

Once you understand the different types of circuit boards and how buyers evaluate them, the next step is learning where they come from. This is one of the most common questions beginners ask after watching videos of experienced electronic recyclers sorting pallets of motherboards, servers, and other electronic equipment. The answer is that successful recyclers rarely depend on luck. Instead, they understand where electronic equipment naturally reaches the end of its useful life and build relationships with people and organizations that regularly replace or dispose of older technology. By focusing on ethical and legal sourcing methods, you can develop a reliable supply of circuit boards while building a positive reputation within the recycling community.

Residential Electronics and Household E-Waste

Many opportunities to find circuit boards begin much closer to home than people realize. Every year, homeowners replace computers, televisions, gaming systems, printers, routers, stereos, kitchen appliances, and countless other electronic devices. While these products may no longer function or may simply be outdated, many still contain valuable circuit boards that can be responsibly recycled.

One of the simplest ways to begin sourcing electronic scrap is by letting friends, family members, neighbors, and coworkers know that you recycle unwanted electronics. Many people store broken computers, outdated networking equipment, old DVD players, and unused electronics in garages, closets, or storage rooms because they are unsure how to dispose of them properly. Offering to responsibly recycle these items can create a steady source of circuit boards while helping others reduce clutter.

Community electronic recycling events also provide valuable opportunities. Many cities and counties organize scheduled collection days where residents can safely dispose of computers, televisions, monitors, printers, and other electronic devices. While collection policies vary by location, these events demonstrate just how much electronic equipment is discarded each year and may introduce you to organizations that specialize in electronics recycling.

Household appliances should not be overlooked either. Microwave ovens, washing machines, refrigerators, dishwashers, coffee makers, air purifiers, and many other appliances contain electronic control boards. Although these boards may not always be among the highest grades, they still contribute to the overall electronic recycling stream and can become worthwhile when collected consistently.

As you gain experience, you'll begin recognizing electronic scrap opportunities during garage sales, neighborhood cleanouts, moving sales, and home renovation projects. Many people are happy to see unwanted electronics recycled responsibly instead of ending up in a landfill.

Business, School, and IT Asset Opportunities

Businesses and institutions often replace electronics on a much larger scale than individual households. Computer upgrades, office relocations, technology refresh cycles, and equipment modernization projects regularly generate large quantities of electronic scrap that must be managed responsibly.

RECYCLE

Many companies replace desktop computers, laptops, servers, networking equipment, and point-of-sale systems every few years as technology evolves. Schools and universities frequently upgrade computer labs, classroom technology, and administrative equipment. Hospitals, manufacturing facilities, laboratories, and government offices also retire electronic equipment as newer systems become available.

This process is commonly known as **IT asset disposition**, often abbreviated as ITAD. Organizations use ITAD providers or electronics recycling companies to securely remove, manage, and recycle outdated technology. While some equipment is refurbished or resold, other devices are dismantled and recycled, creating opportunities for businesses specializing in electronic scrap.

Building professional relationships with local businesses, computer repair shops, managed IT service providers, electronics recyclers, and nonprofit organizations can create consistent sourcing opportunities. Many organizations appreciate working with reliable individuals who understand responsible recycling practices and respect data security requirements.

Computer repair businesses can also become valuable partners. During repair work, technicians often replace damaged motherboards, expansion cards, power supplies, and other components that are no longer repairable. Although each individual repair may produce only a small amount of electronic scrap, those materials accumulate over time.

Auctions, Liquidations, and Other Legal Sources

Some of the most productive sources of circuit boards come from organizations that regularly sell surplus equipment through public sales. Instead of waiting for individual electronic devices to become available, experienced recyclers often purchase larger quantities of equipment from auctions and liquidation events.

Полка
НЕ
ПРОДАЕТСЯ
LOT#
39

Government agencies, schools, universities, corporations, manufacturers, financial institutions, hospitals, and technology companies frequently sell retired electronic equipment through public auction platforms. These auctions may include computers, servers, networking hardware, industrial control systems, telecommunications equipment, laboratory instruments, and many other devices containing valuable circuit boards.

Business liquidations represent another excellent opportunity. When companies relocate, downsize, merge, or close operations, large quantities of office electronics and industrial equipment often become available through liquidation firms. Purchasing these materials legally allows recyclers to acquire inventory while helping businesses responsibly clear unused assets.

Storage unit auctions, estate sales, surplus warehouse sales, and nonprofit donation centers can also produce worthwhile opportunities. While not every purchase will contain valuable circuit boards, experienced buyers learn to recognize equipment categories that consistently provide strong returns.

The most successful electronic recyclers understand that reliable sourcing depends on relationships rather than chance. By consistently acting professionally, honoring agreements, arriving on time, and treating sellers with respect, you build a reputation that encourages people to contact you whenever additional electronic equipment becomes available.

Last Thoughts

Finding circuit boards legally begins with understanding where electronic equipment naturally reaches the end of its useful life. Homes, businesses, schools, IT departments, public auctions, liquidation companies, and electronics recycling programs all generate valuable opportunities for responsible recyclers. By focusing on relationship building, ethical sourcing, and professional conduct, you'll establish a dependable foundation that supports long-term success. In the next chapter, you'll learn how to purchase electronic scrap wisely, organize and prepare circuit boards efficiently, and use the basic tools that help experienced recyclers work safely and productively.

Chapter 3: Buying, Collecting, and Preparing Circuit Boards

After learning where circuit boards come from, the next step is understanding how to acquire them wisely and prepare them for sale. Many beginners assume that every electronic device is worth purchasing or dismantling, but experienced recyclers know that successful buying begins long before money changes hands. Learning how to evaluate equipment, organize your inventory, and prepare circuit boards correctly helps you save time, reduce unnecessary work, and maximize the value of every purchase. This chapter focuses on making informed decisions that allow you to build an efficient and profitable electronic recycling operation.

Buying Electronic Scrap at Auctions and Sales

Purchasing electronic scrap can provide a steady source of inventory when approached carefully. Public auctions, business liquidations, estate sales, surplus warehouses, school equipment sales, and government auctions regularly offer computers, servers, networking equipment, telecommunications hardware, industrial electronics, and other devices containing valuable circuit boards.

D20100
FRAGILE

Before purchasing any electronic equipment, take time to evaluate what is actually being offered. Photographs, equipment lists, manufacturer information, and model numbers often provide clues about the type of circuit boards inside. Older enterprise servers, industrial control systems, telecommunications equipment, and professional networking hardware frequently contain boards that differ significantly from ordinary consumer electronics.

Whenever possible, establish a maximum purchase price before bidding begins. It is easy to become caught up in the excitement of an auction, especially when several buyers recognize valuable equipment. Experienced recyclers remain disciplined by calculating the potential value of the circuit boards, considering transportation costs, labor, storage requirements, and any remaining materials that must be responsibly disposed of after processing.

Estate sales and surplus sales often require a different approach. Instead of buying large quantities of identical equipment, you may encounter individual computers, office electronics, laboratory instruments, networking hardware, or commercial equipment. Learning to quickly recognize valuable categories allows you to make confident purchasing decisions while avoiding unnecessary expenses.

Patience is one of the most valuable skills a buyer can develop. Not every auction will produce worthwhile opportunities, and not every electronic device contains high-value circuit boards. Waiting for the right purchase often produces better long-term results than buying everything available.

Sorting, Organizing, and Storing Circuit Boards

Once electronic equipment has been acquired, organization becomes one of the most important parts of the recycling process. Properly sorting circuit boards makes grading easier, improves communication with buyers, and reduces the possibility of accidentally mixing higher-value boards with lower-grade materials.

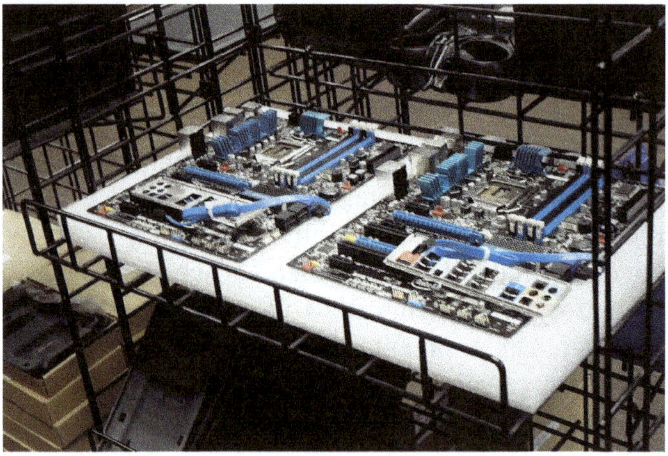

Begin by separating circuit boards according to their general category whenever practical. Computer motherboards, server boards, networking equipment, telecommunications boards, industrial control boards, power supply boards, and appliance control boards should each have their own storage area. Keeping similar materials together makes future grading significantly easier while helping you become more familiar with each category over time.

Organization also protects your inventory. Circuit boards should be stored in clean, dry locations where they are protected from excessive moisture, dirt, and unnecessary physical damage. Although electronic scrap is ultimately being recycled, preserving the condition of the boards helps buyers inspect them more efficiently and may prevent avoidable damage during storage or transportation.

Labeling storage containers is another habit that quickly pays dividends. Clearly marked bins or shelving allow you to identify materials at a glance, simplify inventory management, and reduce the amount of time spent resorting boards before each sale. As your inventory grows, good organization becomes increasingly valuable.

Some recyclers also maintain simple inventory records that include the source of the equipment, approximate quantities, purchase costs, and estimated grades. These records make it easier to evaluate which sourcing methods consistently produce the best results and help guide future purchasing decisions.

Tools Every Beginner Should Know

Unlike many industrial businesses, electronic recycling can begin with a relatively modest collection of tools. Most beginners already own many of the items needed to safely dismantle electronic equipment and organize circuit boards. The goal is not to purchase every available tool but to work efficiently while protecting both yourself and the materials you recover.

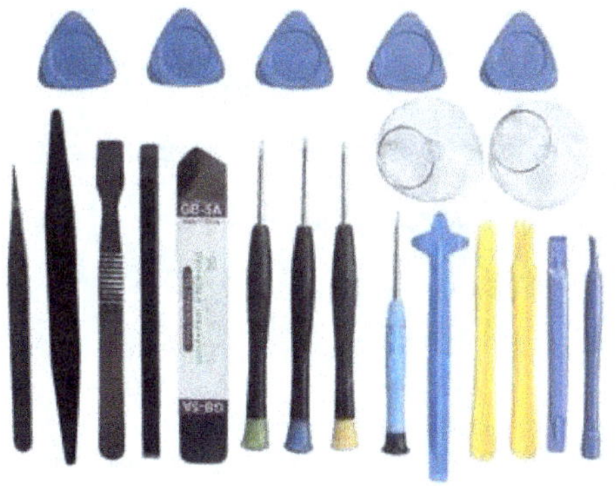

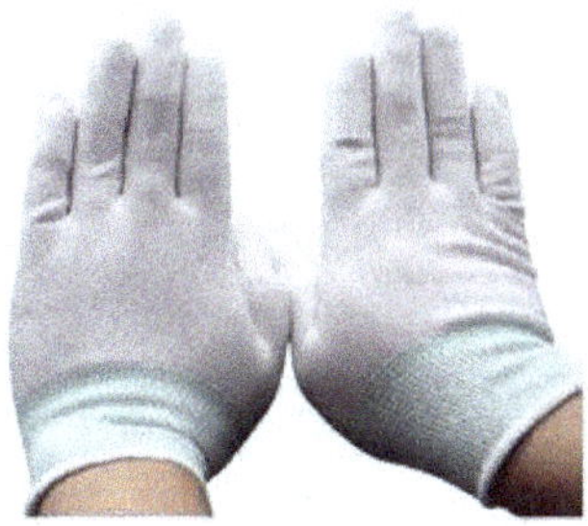

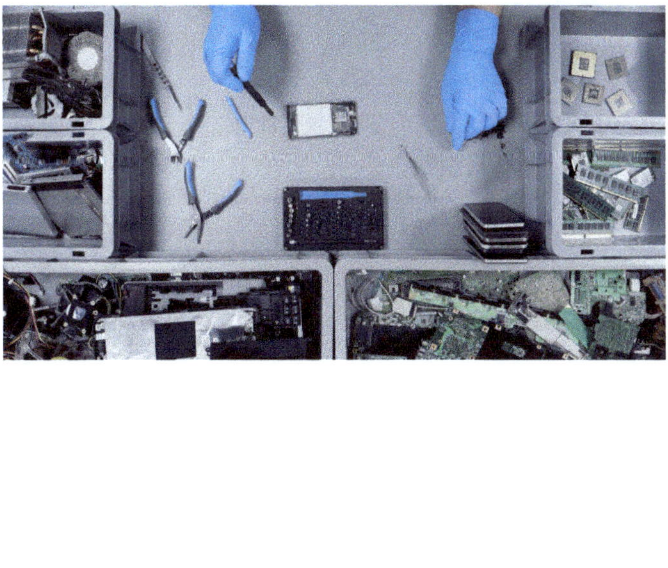

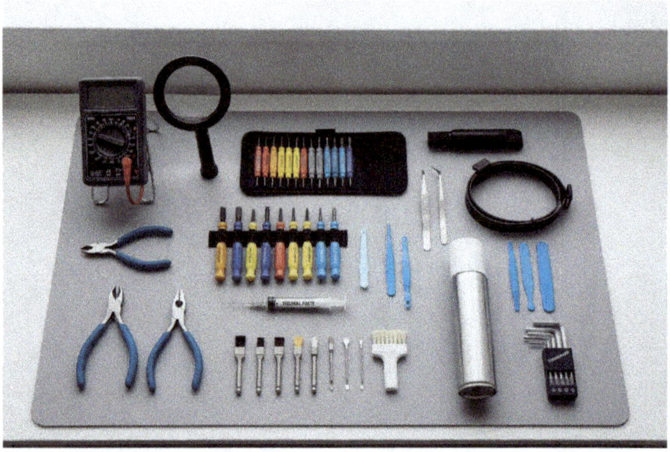

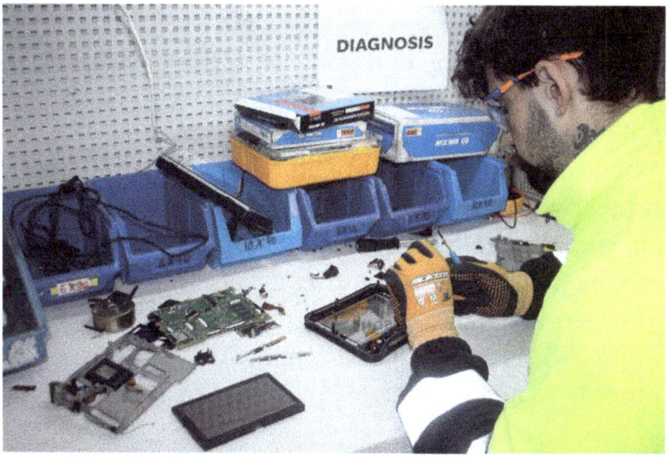

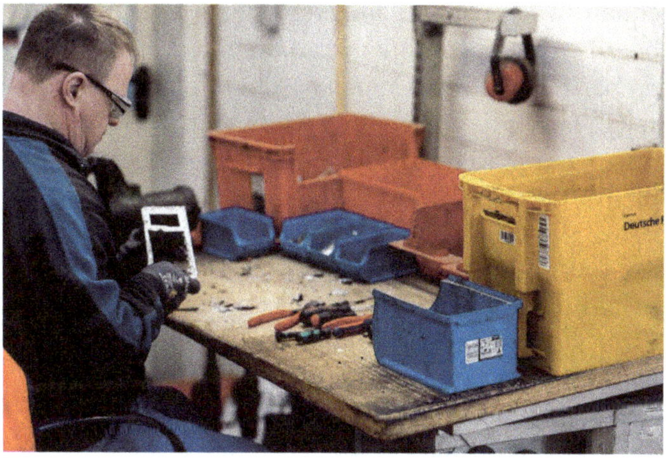

A quality screwdriver set is one of the most important tools because electronic equipment uses a variety of fasteners. Phillips, flathead, Torx, and hex drivers are commonly needed when opening computers, servers, networking equipment, and consumer electronics. Needle-nose pliers, side cutters, and small pry tools are also useful when carefully removing components during disassembly.

Safety equipment should always be considered essential rather than optional. Safety glasses help protect your eyes from broken plastic, flying debris, and sharp metal edges, while durable work gloves reduce the risk of cuts during disassembly. Keeping your work area clean and organized further improves safety while making the dismantling process more efficient.

Storage bins, shelving, and sturdy containers are equally valuable. Organized storage prevents circuit boards from becoming mixed together and makes preparing shipments much easier when it is time to sell. Many experienced recyclers consider organization to be just as important as the tools themselves.

As your operation grows, you may choose to invest in additional equipment such as pallet scales, rolling carts, shelving systems, cordless power drivers, or inventory management software. However, these investments should come only after you understand your workflow and identify genuine needs. Successful recyclers focus first on developing good habits before purchasing specialized equipment.

Last Thoughts

Buying, collecting, and preparing circuit boards requires careful planning rather than impulse decisions. Learning how to evaluate purchases, organize inventory, and use the proper tools creates a more efficient operation while helping you avoid costly mistakes. As your experience grows, these habits will save countless hours and improve the quality of every shipment you prepare. In the next chapter, you'll learn how electronic scrap buyers evaluate circuit boards, why prices change over time, and the practical strategies experienced recyclers use to maximize the value of every sale.

Chapter 4: Selling Circuit Boards for Maximum Value

Once your circuit boards have been sourced, organized, and properly stored, the final step is turning those materials into income. While selling electronic scrap may seem straightforward, experienced recyclers understand that preparation, timing, and knowledge often determine how much a buyer is willing to pay. Knowing how electronic scrap buyers evaluate boards, understanding why prices change, and presenting well-organized material can significantly improve your results over time. This chapter explains how the selling process works and how thoughtful preparation can help you maximize every transaction.

How Electronic Scrap Buyers Evaluate Boards

Electronic scrap buyers purchase circuit boards with the intention of recovering valuable materials or reselling them to larger processing facilities. Because every shipment is different, buyers carefully inspect incoming boards before determining their value. The goal is to estimate the amount of recoverable material while considering the cost of transportation, sorting, processing, and refining.

FRANKFURT
FRANKFURT

One of the first things buyers evaluate is how well the circuit boards have been sorted. Shipments containing clearly separated categories of boards are generally easier to inspect than mixed loads containing many different types of electronic scrap. Keeping server boards, computer motherboards, networking equipment, telecommunications boards, industrial controls, and appliance boards separated allows buyers to evaluate each category more accurately.

Buyers also examine the overall condition of the boards. Boards that remain largely intact are often easier to identify than those that have been heavily damaged during dismantling. Excessive contamination, unnecessary attachments, moisture damage, or large amounts of non-electronic debris may reduce efficiency during processing and affect the overall value of a shipment.

Experience also plays an important role. Professional buyers become familiar with thousands of different circuit board designs over time and often recognize valuable equipment immediately. As you gain experience, you'll begin developing many of the same identification skills, making it easier to estimate the potential value of electronic scrap before you even purchase it.

Building professional relationships with buyers can also be beneficial. Honest communication, accurate representation of your materials, and consistently delivering organized shipments help establish trust while making future transactions smoother and more efficient.

Understanding Circuit Board Prices and Market Trends

Like many recyclable materials, circuit board prices change over time. Beginners are sometimes surprised to discover that the same type of board may sell for different amounts depending on market conditions, buyer demand, and the overall economics of the recycling industry.

Several factors influence pricing. The value of copper, gold, silver, palladium, and other recoverable metals can affect how much electronic scrap buyers are willing to pay. Manufacturing demand, refining capacity, transportation costs, and global commodity markets also contribute to changes in pricing throughout the year.

Electronic technology itself also influences demand. As newer products enter the marketplace, certain categories of older equipment become more common while others become increasingly scarce. Some specialized industrial boards or telecommunications equipment may remain desirable because they contain components or materials that are less common in modern consumer electronics.

Monitoring market trends helps you make more informed decisions about when to sell your inventory. While most small recyclers sell materials regularly rather than attempting to predict market movements, understanding broader trends provides useful context when evaluating offers from different buyers.

It can also be worthwhile to compare prices between multiple electronic scrap buyers. Different companies specialize in different categories of material and may offer different pricing structures depending on their processing methods or customer base. Taking time to understand your local market helps ensure that you're making informed selling decisions.

Maximizing Your Profits

Improving profitability rarely depends on finding one extraordinary circuit board. Instead, it comes from consistently applying good habits throughout the entire sourcing and selling process. Every decision, from how you acquire equipment to how you prepare your inventory, contributes to your overall results.

One of the simplest ways to improve profitability is by maintaining excellent organization. Clearly separating board categories, removing unnecessary debris, and storing materials properly allows buyers to inspect shipments more efficiently while reducing the possibility of lower pricing caused by mixed or contaminated loads.

Transportation costs should also be considered. Fuel, shipping expenses, packaging materials, and travel time all affect your final profit. Combining multiple sales into larger shipments or coordinating deliveries with other business activities may improve efficiency while reducing overall operating costs.

Maintaining accurate records can also become a valuable business tool. Tracking where materials were obtained, how much they cost, what buyers paid, and which sources consistently produced higher-grade boards allows you to identify trends that improve future decision-making. Over time, these records become increasingly valuable as your operation grows.

Perhaps the most important lesson is that patience often produces better results than rushing into every opportunity. Not every electronic device is worth purchasing, not every buyer offers the same value, and not every shipment needs to be sold immediately. Experienced recyclers carefully evaluate each situation, balancing storage capacity, market conditions, and cash flow before making selling decisions.

By consistently focusing on quality, organization, and professionalism, you'll gradually build a reputation that supports long-term success while increasing the efficiency of your electronic recycling operation.

Last Thoughts

Selling circuit boards successfully involves much more than finding a buyer. Understanding how electronic scrap buyers evaluate materials, staying informed about market trends, and consistently presenting organized inventory all contribute to stronger results over time. The knowledge you gain through experience will help you recognize valuable opportunities while avoiding many of the common mistakes beginners make. In the next chapter, you'll learn how to move beyond occasional sales by developing reliable sourcing relationships, avoiding costly business mistakes, and building an ethical electronic recycling side business that can continue growing for years to come.

Chapter 5: Building an Ethical E-Waste Side Business

Many people begin collecting circuit boards simply out of curiosity. They recycle an old computer, purchase a small lot of electronics at an auction, or receive unwanted equipment from family members. As their knowledge grows, they often realize that electronic recycling offers much more than occasional opportunities. With organization, consistency, and ethical business practices, it can become a reliable side business that generates ongoing income while helping reduce electronic waste. Like any successful business, however, long-term success depends on relationships, professionalism, and continuous learning rather than chasing quick profits.

Developing Reliable Sources

The most successful electronic recyclers rarely depend on finding random computers sitting beside the road or hoping someone gives them unwanted electronics. Instead, they focus on building dependable relationships with organizations that regularly replace electronic equipment.

B

Businesses are among the best long-term sources of circuit boards because technology is constantly evolving. Companies replace desktop computers, laptops, servers, networking hardware, point-of-sale systems, and office electronics on a regular schedule. By introducing yourself professionally and explaining that you responsibly recycle electronic equipment, you may become someone they contact whenever equipment is retired.

Computer repair shops can also become valuable partners. Every week they replace damaged motherboards, graphics cards, power supplies, networking equipment, and other electronic components that are no longer economical to repair. While individual repairs may generate only a small amount of electronic scrap, those materials quickly accumulate over time.

Schools, universities, nonprofit organizations, libraries, and local government offices also periodically upgrade their technology. Many of these organizations participate in surplus sales or electronics recycling programs, while others work directly with approved recyclers. Understanding how these organizations manage retired equipment can help you identify additional sourcing opportunities.

Consistency is what transforms one-time opportunities into long-term relationships. Arriving when promised, communicating professionally, handling equipment responsibly, and respecting data privacy demonstrate that you are someone others can trust. Over time, that reputation often becomes your greatest competitive advantage.

Avoiding Common Beginner Mistakes

Every recycler makes mistakes while learning, but recognizing common problems early can save significant time, money, and frustration. Many beginners become excited by the volume of electronics they find without carefully evaluating whether those items actually contain valuable circuit boards or justify the effort required to process them.

IT EQUIPMENT
SMALL APPLIANCES

SCRAP ONLY
MIXED
LOT
For SCRAP / PARTS

One of the most common mistakes is collecting every electronic device simply because it appears old. Age alone does not determine value. Some newer equipment may contain higher-value boards than much older devices, while certain products contain very little recoverable material regardless of when they were manufactured. Learning to identify worthwhile equipment before transporting it saves both time and storage space.

Another mistake involves poor organization. Allowing server boards, appliance boards, networking equipment, computer motherboards, expansion cards, and miscellaneous electronics to become mixed together makes grading more difficult and often reduces efficiency when preparing shipments. Developing organized storage habits from the beginning simplifies every stage of the recycling process.

Some beginners also purchase expensive equipment too early. Large shelving systems, pallet jacks, warehouse space, and specialized tools may eventually become worthwhile investments, but they are rarely necessary during the early stages. Building experience first allows you to make smarter purchasing decisions later as your business grows.

Perhaps the most important mistake to avoid is overlooking data security. Computers, servers, networking equipment, and storage devices may still contain sensitive information when they enter the recycling stream. Hard drives, solid-state drives, backup media, and other storage devices should always be handled responsibly and in accordance with applicable laws, organizational policies, and customer expectations. Respecting data privacy builds trust and strengthens your professional reputation.

Scaling Your Circuit Board Recycling Business

As your knowledge and confidence increase, you may begin expanding beyond occasional electronic recycling projects into a more structured operation. Growth does not necessarily mean processing dramatically larger quantities of equipment. More often, it results from improving efficiency, strengthening sourcing relationships, and developing better systems for managing inventory.

METAL
PLASTIC
BATTERY

EN UND AKKUS
EINWERFEN!
SAMMELBOX

COMPUTATION.CA | TORONTO
COMPONENTS
HARD DRIVES
PLASTICS
METALS
PROUD PARTICIPANT, EARTH WEEK TORONTO

Organization becomes increasingly valuable as inventory grows. Clearly labeled storage areas, dedicated workspaces, inventory tracking, and scheduled pickup or purchasing routines help create a smooth workflow while reducing unnecessary handling. Instead of constantly sorting mixed electronics, you'll spend more time evaluating, grading, and preparing materials for sale.

Diversifying your sourcing methods also creates greater stability. Rather than depending on a single business or one auction platform, experienced recyclers often build relationships with several different sources. Computer repair shops, businesses, schools, IT departments, liquidation companies, auctions, electronics recyclers, and community organizations all contribute to a more consistent supply of electronic scrap throughout the year.

Keeping accurate records also becomes increasingly important as your operation expands. Tracking purchases, sales, transportation costs, inventory levels, and sourcing locations helps you identify which activities produce the greatest return. These records provide valuable information that supports smarter business decisions while making it easier to improve efficiency over time.

Most importantly, continue protecting your reputation. The electronic recycling industry is built on trust. Businesses want confidence that their equipment will be handled responsibly. Buyers appreciate honest grading and organized shipments. Suppliers prefer working with dependable individuals who communicate professionally and honor their commitments. A strong reputation creates opportunities that cannot be purchased and often becomes one of the most valuable assets your business possesses.

Last Thoughts

Building an ethical e-waste side business is a gradual process built on knowledge, relationships, organization, and integrity. Every successful recycler begins with small opportunities and develops stronger systems through experience and continuous learning. By focusing on dependable sourcing, avoiding common mistakes, and expanding thoughtfully, you'll build a business that can continue growing alongside the rapidly evolving electronics industry. In the next chapter, you'll explore advanced opportunities in electronic recycling, discover valuable materials beyond circuit boards, and learn how the broader recycling economy creates new possibilities for long-term success.

Chapter 6: Advanced Opportunities in Electronic Recycling

As your experience in electronic recycling grows, you'll begin recognizing opportunities that extend well beyond individual circuit boards. What may have started as a simple interest in computer motherboards can gradually develop into a broader understanding of how valuable materials move through the modern economy. Experienced recyclers learn to evaluate complete systems rather than individual components, identify emerging opportunities, and adapt as technology continues to evolve. This final chapter explores how expanding your knowledge can help you continue growing while remaining committed to ethical recycling and responsible resource management.

Beyond Circuit Boards: Other Valuable Electronic Scrap

Although circuit boards are often the primary focus of electronic recyclers, they represent only one part of a much larger industry. Modern electronic devices contain a wide variety of recyclable materials that can create additional opportunities once you understand where to look.

Power supplies, computer memory modules, graphics cards, processors, hard drives, solid-state drives, networking equipment, telecommunications hardware, industrial controllers, electric motors, copper wiring, aluminum heat sinks, stainless steel enclosures, and brass connectors are just a few examples of materials that commonly accompany circuit boards. While some components may have resale value for reuse, others are sold through specialized recycling channels that recover their raw materials.

Learning to recognize these additional materials helps you maximize the value of each electronic device you acquire. Instead of viewing an old desktop computer as a single item, experienced recyclers often evaluate the motherboard, processor, memory, power supply, aluminum components, copper wiring, steel case, and other materials separately. This broader perspective creates more opportunities while reducing unnecessary waste.

It is equally important to understand that not every component should be dismantled immediately. Some equipment, particularly newer computers, networking hardware, industrial electronics, and specialty devices, may have greater value when sold intact for refurbishment or reuse rather than being dismantled for scrap. Taking time to evaluate each item before processing allows you to choose the option that provides the greatest overall value.

As technology continues to change, new categories of electronic equipment will enter the recycling stream. Remaining curious and willing to learn ensures that you'll continue recognizing opportunities as industries adopt new devices and retire older ones.

Understanding the Circular Economy

Electronic recycling is much more than recovering valuable materials. It plays an important role in supporting the **circular economy**, where products and raw materials remain in use for as long as possible instead of being discarded after a single use.

Every year, manufacturers produce millions of electronic devices that require copper, aluminum, gold, silver, plastics, rare earth elements, and many other raw materials. Recovering these materials through responsible recycling helps reduce the need for additional mining while conserving energy and minimizing waste.

Many electronic devices also follow several stages before being recycled. Equipment may first be repaired, refurbished, upgraded, donated, or resold before eventually entering the recycling stream. Understanding this process helps recyclers appreciate that their role extends beyond simply selling circuit boards. They contribute to a larger system that supports responsible resource management and environmental sustainability.

Businesses, governments, schools, manufacturers, and consumers all participate in this cycle. As technology continues advancing, the volume of electronic equipment reaching the end of its useful life will likely continue increasing, making responsible recycling an increasingly important part of the global economy.

Viewing electronic recycling through this broader perspective also reinforces the importance of ethical sourcing. Recovering materials responsibly protects valuable resources while supporting industries that depend upon recycled materials to manufacture tomorrow's products.

Long-Term Success in the E-Waste Industry

Long-term success in electronic recycling is rarely the result of a single profitable purchase. Instead, it develops through continuous learning, professional relationships, and a willingness to adapt as technology changes.

Technology evolves quickly. Computers become smaller, communication systems become faster, manufacturing equipment becomes more advanced, and entirely new categories of electronic devices appear every year. Successful recyclers recognize these changes as opportunities to continue learning rather than obstacles to overcome.

Developing strong relationships remains equally important. Businesses, schools, repair shops, IT departments, auction companies, liquidation firms, electronics recyclers, and professional buyers all contribute to the electronic recycling ecosystem. Treating every interaction professionally helps establish trust while creating opportunities that may continue for many years.

Maintaining organized records also supports long-term growth. Tracking inventory, purchases, sales, transportation costs, buyer preferences, and sourcing locations provides valuable information that helps improve future decision-making. Small improvements made consistently often produce significant results over time.

Perhaps the greatest advantage experienced recyclers possess is curiosity. They continue asking questions, studying unfamiliar equipment, learning new grading systems, and exploring emerging areas of the recycling industry. That commitment to continuous education allows them to recognize opportunities long before they become widely known.

No matter how large your operation becomes, remember that integrity remains your most valuable asset. Honest communication, ethical sourcing, responsible recycling practices, and respect for both customers and suppliers build a reputation that supports lasting success. Markets may change, technologies may evolve, and prices may fluctuate, but professionalism and trust remain valuable in every stage of your journey.

Last Thoughts

Electronic recycling is a constantly evolving industry filled with opportunities for individuals who are willing to learn, adapt, and operate responsibly. By expanding your knowledge beyond circuit boards, understanding the role of the circular economy, and committing yourself to continuous improvement, you'll be well prepared to grow alongside this rapidly changing field. Every computer, server, networking device, industrial controller, and electronic product represents more than discarded technology. It represents an opportunity to recover valuable resources, reduce waste, support sustainable manufacturing, and build a rewarding business based on knowledge, integrity, and responsible recycling.

Last Thoughts

Every successful electronic recycler begins with curiosity. Few people start with warehouses full of equipment or long lists of business contacts. Instead, they begin by learning how electronic devices are built, understanding which components have value, and discovering how those materials move through the recycling industry. As their knowledge grows, so do the opportunities around them. What once looked like outdated computers or discarded networking equipment gradually becomes a source of valuable resources that can be recovered responsibly and returned to the manufacturing economy.

Throughout this book, you've seen that success in electronic recycling depends on much more than simply collecting circuit boards. It begins with understanding what makes different boards valuable, learning how buyers grade them, identifying legitimate sourcing opportunities, and preparing materials in ways that improve both efficiency and profitability. Every stage of the process contributes to the final result, and each skill you develop strengthens your ability to recognize worthwhile opportunities in the future.

One of the most valuable lessons is that ethical sourcing creates lasting success. Businesses, schools, government agencies, homeowners, and recycling companies all generate electronic waste through normal technology replacement cycles. By focusing on these legitimate sources, respecting property rights, protecting sensitive information, and maintaining professional relationships, you build a reputation that opens doors to opportunities far beyond a single transaction.

BUSINESS-TO-BUSINESS
ELECTRONICS RECYCLING SOLUTIONS

Electronic recycling is also an industry built on continuous learning. Technology changes every year. New computer systems replace older models, communication equipment evolves, industrial machinery becomes more advanced, and entirely new categories of electronic devices enter the marketplace. These changes constantly create new sources of electronic scrap while introducing new types of circuit boards and recyclable materials. Remaining curious and willing to continue learning allows you to grow alongside the industry instead of being left behind by it.

As your experience increases, you may also discover opportunities beyond selling circuit boards. Some recyclers expand into complete electronic asset recovery, while others specialize in business cleanouts, IT asset disposition, industrial surplus purchasing, computer refurbishment, or broader electronic recycling services. The knowledge you've developed throughout this guide provides a strong foundation that can support many different paths depending on your interests and goals.

Building a successful electronic recycling business is rarely about making one extraordinary sale. More often, it is the result of making hundreds of good decisions over time. Choosing the right equipment to purchase, organizing inventory carefully, building dependable relationships, comparing buyers, maintaining accurate records, and treating every customer professionally all contribute to long-term growth. Small improvements made consistently often produce the greatest results.

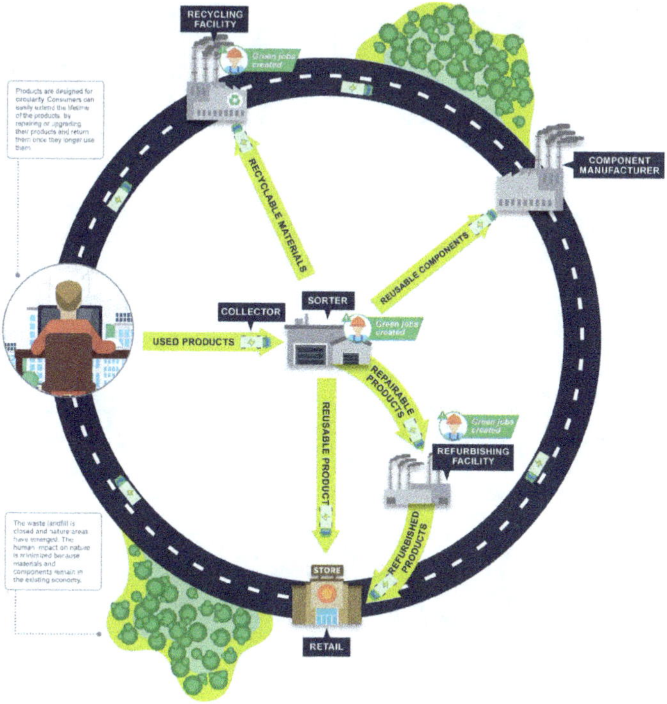
RECYCLING FACILITY
Green jobs created
Products are designed for circularity. Consumers can easily extend the lifetime of the products by repairing or upgrading their products and return them once they longer use them.
RECYCLABLE MATERIALS
COMPONENT MANUFACTURER
REUSABLE COMPONENTS
SORTER
COLLECTOR
Green jobs created
USED PRODUCTS
REPAIRABLE PRODUCTS
Green jobs created
REFURBISHING FACILITY
REUSABLE PRODUCT
The waste landfill is closed and nature areas have emerged. The human impact on nature is minimized because materials and components remain in the existing economy.
REFURBISHED PRODUCTS
STORE
RETAIL

Finally, remember that every circuit board responsibly recycled represents more than an opportunity to earn income. It also helps recover valuable resources, reduce electronic waste, conserve natural materials, and support manufacturers that rely on recycled metals to produce tomorrow's technology. Whether your goal is developing a rewarding hobby, earning supplemental income, or building a growing side business, your work contributes to a larger system that benefits both the economy and the environment.

Thank you for reading **Selling Circuit Boards for Profit Guide**. I hope this book has provided the practical knowledge and confidence needed to begin your journey into electronic recycling. Continue asking questions, continue expanding your knowledge, and continue approaching every opportunity with professionalism, integrity, and respect. The more you learn, the more you'll discover that the greatest value in electronic recycling comes not only from the materials themselves, but from the knowledge that allows you to recognize opportunities that others often overlook.

Image Sources, References, and Additional Resources

Public Domain Image Sources

The public domain images used throughout this book were obtained or adapted from the following trusted repositories. These collections provide educational, historical, industrial, and government-produced images that are generally suitable for educational publishing. Always verify the licensing status of individual images before publication.

- Library of Congress Prints & Photographs Division
  https://www.loc.gov/pictures/

- U.S. National Archives
  https://www.archives.gov/research

- U.S. Department of Energy Image Gallery
  https://www.energy.gov/

- National Institute of Standards and Technology (NIST) Image Gallery
  https://www.nist.gov/

- U.S. Environmental Protection Agency (EPA) Multimedia Resources https://www.epa.gov/

- U.S. Geological Survey (USGS) Multimedia Gallery https://www.usgs.gov/media/images

- NASA Image and Video Library https://images.nasa.gov/

- Wikimedia Commons (Public Domain and Public Domain Marked Images) https://commons.wikimedia.org/

- Rawpixel Public Domain Collection https://www.rawpixel.com/category/53/public-domain

- Openverse https://openverse.org/

References

Basel Convention. *Technical Guidelines on the Environmentally Sound Management of Waste Electrical and Electronic Equipment (WEEE).*

Institute of Scrap Recycling Industries (ISRI). *Electronics Recycling Resources.*

U.S. Environmental Protection Agency (EPA). *Electronics Donation and Recycling.*

U.S. Geological Survey (USGS). *Mineral Commodity Summaries*.

National Institute of Standards and Technology (NIST). *Electronics Manufacturing and Standards*.

United Nations Institute for Training and Research (UNITAR). *Global E-Waste Monitor.*

International Telecommunication Union (ITU). *Global E-Waste Statistics*.

Occupational Safety and Health Administration (OSHA). *Electrical Safety and Workplace Safety Resources*.

National Institute for Occupational Safety and Health (NIOSH). *Workplace Safety Publications*.

Additional Resources

Institute of Scrap Recycling Industries (ISRI)
https://www.isri.org/

U.S. Environmental Protection Agency, Electronics Recycling
https://www.epa.gov/recycle/electronics-donation-and-recycling

United Nations Institute for Training and Research (UNITAR)
https://unitar.org/

Global E-Waste Statistics Partnership
https://globalewaste.org/

U.S. Geological Survey (USGS)
https://www.usgs.gov/

National Institute of Standards and Technology (NIST)
https://www.nist.gov/

Occupational Safety and Health Administration (OSHA)
https://www.osha.gov/

Sustainable Electronics Recycling International (SERI)
https://sustainableelectronics.org/

Electronics TakeBack Coalition
https://www.electronicstakeback.com/

About the Author

Roggie “Raj” Clark is an entrepreneur and a 10-year marketing professional with a proven track record in e-commerce and digital business growth. Having built and managed successful online stores with thousands of sales and consulted for numerous marketing clients, Roggie has mastered the intricacies of online marketing, branding, and consumer behavior. He has also written several popular books on digital marketing, SEO (search engine optimization), and programming (HTML, CSS, and JavaScript).

As a former marketing manager at a fintech company—a role he secured through his own marketing and financial expertise—Roggie gained firsthand experience in financial strategy and wealth-building. This journey inspired him to share insights with amateur investors, helping them save money, make smarter financial decisions, and build long-term wealth. Raj currently lives in Maryland with his sister and niece.

www.ingramcontent.com/pod-product-compliance
Lightning Source LLC
LaVergne TN
LVHW050824061126
851857LV00012B/959